Exploring *The*
BUILDING BLOCKS
of
Science

Book 5
TEACHER'S MANUAL

REBECCA W. KELLER, PhD

Exploring the Building Blocks of Science Book 5 Teacher's Manual
ISBN 978-1-941181-11-9

Published by Gravitas Publications Inc.
www.realscience4kids.com
www.gravitaspublications.com

A Note From the Author

This curriculum is designed for middle school level students and provides further exploration of the scientific disciplines of chemistry, biology, physics, geology, and astronomy. *Exploring the Building Blocks of Science Book 5 Laboratory Notebook* accompanies *Exploring the Building Blocks of Science Book 5 Student Textbook*. Together, both provide students with basic science concepts needed for developing a solid framework for real science investigation. The *Laboratory Notebook* contains 44 experiments—two experiments for each chapter of the Student Textbook. These experiments allow students to further explore concepts presented in the *Student Textbook*. This *Teacher's Manual* will help you guide students through laboratory experiments designed to help students develop the skills needed for using the scientific method.

There are several sections in each chapter of the *Laboratory Notebook*. The section called *Think About It* provides questions to help students develop critical thinking skills and spark their imagination. The *Experiment* section provides students with a framework to explore concepts presented in the *Student Textbook*. In the *Conclusions* section students draw conclusions from the observations they have made during the experiment. A section called *Why?* provides a short explanation of what students may or may not have observed. And finally, in each chapter an additional experiment is presented in *Just For Fun*.

The experiments take up to 1 hour. The materials needed for each experiment are listed on the following pages and also at the beginning of each experiment.

Enjoy!

Rebecca W. Keller, PhD

Materials at a Glance

Experiment 1	Experiment 3	Experiment 4	Experiment 5	Experiment 6

family photos
birth certificate
other family documents
	food labels, several (students' choice)
	periodic table of elements from *Student Textbook*
	resources (books or online) such as:
	  dictionary
	  encyclopedia
	
	Optional
	computer with internet access
		small, colored marshmallows, 1 pkg
		large marshmallows, 1 pkg
		(could also use gumdrops and/or jellybeans in place of marshmallows)
		toothpicks, 1 box
		
		Optional
		food coloring
			baking soda
			lemon juice
			balsamic vinegar
			salt and water:
			  15-30 ml salt dissolved in 120 ml water
			  (1-2 tbsp. salt dissolved in 1/2 cup of water)
			2 or more egg whites
			milk
			small jars, 7 or more
			measuring cups and spoons
			eye dropper
			
			Peanut Brittle
			360 ml (1 1/2 cups) sugar
			240 ml (1 cup) white corn syrup
			120 ml (1/2 cup) water
			360 ml (1 1/2 cups) raw peanuts (can be omitted)
			5 ml (1 teaspoon) baking soda
			buttered pan
				pencil and eraser
				
				Objects chosen by students, such as:
				  rubber ball
				  cotton ball
				  orange
				  banana
				  apple
				  paper
				  sticks
				  leaves
				  rocks
				  grass
				  Legos
				  building blocks
				  other objects
				
				Optional
				several sheets of paper

Experiment 2

imagination

Experiment 7	Experiment 8	Experiment 9	Experiment 10	Experiment 11
tincture of iodine [**VERY POISONOUS—DO NOT ALLOW STUDENTS TO EAT** any food items that have iodine on them] bread, 1 slice timer wax paper marking pen cup refrigerator green vegetable, 1 one or more other vegetables	pencil colored pencils or crayons student-selected materials to build a model cell	dehydrated agar powder* distilled water cooking pot measuring spoons measuring cup cup plastic petri dishes (20)** cotton swabs permanent marker oven mitt or pot holder	tennis ball paperclip yarn or string (about 3 meters [10 ft]) marble bouncing ball, 1 (or 2 or more of different sizes) **Optional** penknife, ice pick, awl, or other sharp tool pliers	Slinky several paperclips 1-2 apples 1-2 lemons or limes 1-2 oranges 1-2 bananas spring balance scale or food scale meterstick, yardstick, or tape measure tape

* http://www.hometrainingtools.com/nutrient-agar-8-g-dehydrated/p/CH-AGARN08/
** A stack of 20 can be ordered from: http://www.hometrainingtools.com/petri-dishes-plastic-20-pk/p/BE-PETRI20/

Experiment 12	Experiment 13	Experiment 14	Experiment 15	Experiment 16

small to medium size toy car
stiff cardboard
wooden board, smooth and straight (more than 1 meter [3 feet] long)
straight pin or tack, several
small scale or balance
one banana, sliced
10 pennies
meterstick, yardstick or tape measure
tape
	student-selected materials
	several sheets of paper
		pencil, pen, colored pencils
		small jar
		trowel or spoon
		
		Optional
		binoculars
			known mineral samples:
			  calcite
			  feldspar
			  quartz
			  hematite
			several rocks from backyard or nearby
			copper penny
			steel nail
			streak plate (unglazed white ceramic tile)
			paper
			scissors
			marking pen
			tape
			vinegar (small amount)
			lemon juice (small amount)
			eyedropper or spoon
			
			* Find minerals at a local rock and mineral store or order online.
				Students will select materials and use them to make a model of Earth's layers
				
				See Experiment 16 for ingredients for chocolate lava cake

Experiment 17	Experiment 18	Experiment 19	Experiment 21	Experiment 22

student-made brittle candy (see first page of experiment)
1 jar smooth peanut butter (for students with allergies to peanuts, whipped cream can be substituted)
118 ml (1/2 cup) crushed graham crackers
plate or second cookie sheet
materials to make a model volcano of student's choice
	pencil
	flashlight
	compass
	
	A clear night sky away from bright lights is needed.
		basketball
		ping-pong ball
		flashlight
		empty toilet paper tube
		tape
		scissors
		a dark room
		student-selected objects
			modeling clay in the following colors:
			  gray
			  white
			  brown
			  red
			  blue
			  green
			  orange
			butter knife or sculptor's knife
			colored pencils
				a video recording device (camcorder, iPad, cell phone)

Experiment 20

modeling clay in the following colors
  gray
  white
  brown
  red
butter knife or sculptor's knife
ruler
 	
Materials other than clay can be used, such as Styrofoam balls or plaster of Paris and paint.

* A Mineral Scale of Hardness Set of Minerals is available from Home Science Tools: http://www.hometrainingtools.com

Materials
Quantities Needed for All Experiments

Equipment	Foods	Foods (continued)
baking pan ball, bouncing, 1 (or 2 or more of different sizes) ball, ping-pong ball, rubber ball, tennis basketball car, toy, small to medium size compass computer with internet access cookie sheet or plate cup, 1 eye dropper flashlight jars, small, 7 or more knife, butter or sculptor's knife Legos marble, 1 measuring cups and spoons meterstick, yardstick, or tape measure nail, steel oven mitt or pot holder pennies, copper, 10 pot, cooking refrigerator ruler scale, spring balance or food scale scissors Slinky stove streak plate (unglazed white ceramic tile) timer trowel (garden) or spoon video recording device (camcorder, iPad, cell phone) **Optional** binoculars penknife, ice pick, awl, or other sharp tool pliers	apples, 2-3 baking soda bananas, 3-4 bread, 1 slice brittle candy, student-made (see Experiment 17) corn syrup, white, 240 ml (1 cup) egg whites, 2 or more graham crackers, crushed, 118 ml (1/2 cup) lemon juice lemons or limes, 1-2 oranges, 2-3 marshmallows, small, colored, 1 pkg marshmallows, large, 1 pkg (could also use gumdrops and/or jellybeans in place of marshmallows) milk peanuts, raw, 360 ml (1 1/2 cups) (can be omitted) peanut butter, 1 jar smooth, (for students with allergies to peanuts, whipped cream can be substituted) salt, 15-30 ml (1-2 tbsp.) sugar, 360 ml (1 1/2 cups) vegetables, 1 green and students' choice vinegar vinegar, balsamic water See Experiment 16 for ingredients needed for chocolate lava cake. **Optional** food coloring	

Materials
Quantities Needed for All Experiments

Materials	Materials (continued)	Other

agar powder* (dehydrated)
birth certificate
board,wooden, smooth and straight (more than 1 meter [3 feet] long)
cardboard, stiff
clay, modeling in the following colors
    blue
    brown
    gray
    green
    orange
    red
    white
cotton ball, several
cotton swabs
documents, family
eraser
food labels, several (students' choice)
grass
iodine, tincture of [**VERY POISONOUS—DO NOT ALLOW STUDENTS TO EAT** any food items that have iodine on them]
leaves
mineral samples**:
    calcite
    feldspar
    hematite
    quartz
    (Find minerals at a local rock and mineral store or order online.)
misc. model building materials - students' choice
objects, misc. student-chosen
	paper, several sheets
	paperclips, several
	pen, marking
	pencil
	pencils, colored or crayons
	periodic table of elements from *Student Textbook*
	petri dishes, plastic (20)***
	photos, family
	rocks, several
	sticks
	straight pin or tack, several
	tape
	toilet paper tube, empty
	toothpicks, 1 box
	water, distilled
	wax paper
	yarn or string (about 3 meters [10 ft])
		clear night sky away from bright lights
		a dark room
		resources (books or online) such as:
		    dictionary
		    encyclopedia

* http://www.hometrainingtools.com/nutrient-agar-8-g-dehydrated/p/CH-AGARN08/

** A Mineral Scale of Hardness Set of Minerals is available from Home Science Tools: http://www.hometrainingtools.com

*** A stack of 20 can be ordered from: http://www.hometrainingtools.com/petri-dishes-plastic-20-pk/p/BE-PETRI20/

Contents

CHEMISTRY

BIOLOGY

PHYSICS

GEOLOGY

ASTRONOMY

Experiment 1

Writing History

Materials Needed

- family photos
- birth certificate
- other family documents

Objectives

In this experiment students will explore history and how it applies to science.

The objectives of this lesson are for students to:

- Explore history to learn about what has led to the present conditions.
- Understand that science has a history.

Experiment

I. Think About It

Read this section of the *Laboratory Notebook* with your students.

Explore open inquiry with questions such as the following.

- *Do you think you have a history? Why or why not?*

- *Do you think your family has a history? Why or why not?*

- *Do you think knowing your history can be helpful to you? Why or why not?*

- *Do you think science has a history? Why or why not?*

- *Do you think it is important to know the history of scientific discoveries? Why or why not?*

II. Experiment 1: Writing History

Read this section of the *Laboratory Notebook* with your students.

Have the students read the entire experiment.

Objective: An objective is provided.
Hypothesis: A hypothesis is provided.

EXPERIMENT

❶ In the space provided, have the students write a short history of their life to include at least four or five events from birth to present.

❷ Have the students list some resources they could use to reconstruct a more detailed history.

❸-❹ Have the students list resources they could use to reconstruct the histories of their grandparents and great-great-great grandparents.

❺   Have the students pick another family member whose history they can write. Have them gather and list the resources they will use.

❻   In the space provided, have the students use the resources they collected to write the family member's history.

III. Conclusions

Have the students review the results they recorded for the experiment. Have them draw conclusions based on the data they collected.

IV. Why?

Read this section of the *Laboratory Notebook* with your students.
Discuss any questions that might come up.

V. Just For Fun

Have the students think of a technological discovery that helped move science forward. Have them research the development of this technology and write a brief history of it.

Learning to Argue Scientifically

Materials Needed

- imagination

Objectives

In this experiment students will explore thought experiments and presenting scientific arguments.

The objectives of this lesson are for students to:

- Explore thought experiments.
- Explore the development of scientific arguments.

Experiment

I. Think About It

Read this section of the *Laboratory Notebook* with your students.

Ask questions such as the following to guide open inquiry.

- *Do you think it is important for scientists to have arguments about their theories? Why or why not?*

- *Do you think arguing helps scientists better understand science? Why or why not?*

- *Do you think it is helpful for scientists to practice arguing? Why or why not?*

- *Do you think an experiment can be performed by thinking about it? Why or why not?*

- *What do you think a scientist needs to think about before performing an experiment?*

- *What do you think a scientist might need to think about after an experiment has been completed?*

II. Experiment 2: Learning to Argue Scientifically—A Thought Experiment

A thought experiment is done by thinking scientifically about how something might work without actually doing an experiment. In this experiment students will read a fictional play to gain an understanding of how scientists argue their theories, and they will begin to learn how to form the basis of a scientific argument by thinking about it.

Have the students read the entire experiment.

Objective: An objective is provided.
Hypothesis: A hypothesis is provided.

CHEMISTRY

EXPERIMENT

❶ Have the students read the play *The Mystery of Substance: A Philosophical Play* by D. R. Megill.

❷-❹ Have the students answer the questions about the play. There are no "right" answers.

Results

❶-❷ Have the students answer the questions based on what they learned from the play. There are no "right" answers.

III. Conclusions

Have the students use their observations to draw conclusions about arguing scientifically.

IV. Why?

Read this section of the *Laboratory Notebook* with your students.
Discuss any questions that might come up.

V. Just For Fun

Students are to imagine they are on a newly discovered planet that appears to be made entirely of candy. They are to think of experiments to do to prove or disprove this theory. There are no right answers to this thought experiment.

Experiment 3

What Is It Made Of?

Materials Needed

- food labels, several (students' choice)
- periodic table of elements from *Student Textbook*
- resources (books or online) such as:
  dictionary
  encyclopedia

Optional

- computer with internet access

Objectives

In this experiment students will be introduced to the concept that all things are made of atoms and will begin to explore the periodic table of elements.

The objectives of this lesson are for students to:

- Understand that atoms, or elements, are the fundamental components of all things.
- Discover that each type of atom has specific properties.

Experiment

I. Think About It

Read this section of the *Laboratory Notebook* with your students.

Ask questions such as the following to guide open inquiry.

- *Do you think all atoms are the same? Why or why not?*
- *Do you think some atoms are the same? Why or why not?*
- *Do you think it is possible to tell one atom from another? How would you do it?*
- *Do you think you can find out what properties a particular atom has? Why or why not?*
- *How would you find out what your food is made of?*
- *How would you find out what other things are made of?*

II. Experiment 3: What Is It Made Of?

Have the students read the entire experiment.

Objective: An objective is provided for this experiment:
Hypothesis: Have the students write a hypothesis. Some examples:

- *Food labels can be used to tell what is in food.*
- *I can find out what things are made of.*
- *I can use the periodic table of elements to tell me about atoms.*

EXPERIMENT

❶ Answers to the questions:

A. Protons in aluminum: 13
Electrons in aluminum: 13

B. Symbol for carbon: C

C. The elements that have chemical properties similar to helium are neon, argon, krypton, xenon, and radon.

Elements that have the same chemical properties as helium are in the same column in the periodic table.

D. Atomic weight of nitrogen: 14.0067
Number of neutrons in nitrogen: 7

❷ A table is provided for students to record information they discover about the makeup of items of their choice.

The goals of this experiment are to help students begin to investigate the things in their world and to have them start to examine what those things are made of.

There are many possible answers for this experiment. Students will begin to think about what substances are made of and how they are produced. By using basic resources such as the dictionary or encyclopedia, they may not be able to find the elemental composition of all the items they think of.

Some examples of answers are the following:

Things made of metals:

- soda cans and aluminum foil - aluminum
- silverware (steel) - iron, nickel, silver
- coins - copper, nickel
- jewelry - gold, silver

Things we eat:

- salt - sodium and chlorine
- sugar - carbon, oxygen, hydrogen
- water - hydrogen and oxygen
- bread (carbohydrates) - carbon, oxygen, hydrogen, other proteins, and other substances

Also, students can select food items with labels, such as cake mixes, cereal, noodles, and vitamins (with vitamins the label is very detailed so students can also find out how much of something is in the vitamin).

Students DO NOT need to find every component for each item. To say that a cake mix contains salt, flour, and sugar is enough. Let the students go as far as they want to with a particular item. Also, it is not necessary to look up components for each item the students think of. Have them pick a few items they are interested in researching and go from there.

Some examples of information that may be gathered:

ITEM	COMPOSITION	SOURCE
graham crackers	sodium bicarbonate (sodium)	food label
graham crackers	salt (sodium, chlorine)	food label, dictionary - page 1600
car tires	rubber (carbon and hydrogen)	Wikipedia (or www.wikipedia.org)

Results

Students will describe what they discovered about the composition of the items they researched.

Help the students write accurate statements about the data they have collected. Some examples:

- Kellogg's Sugar Smacks™ cereal contains vitamin C, which is called sodium ascorbate.
- Table salt is made of sodium and chlorine.
- Iodized table salt contains sodium, chlorine, and iodine.
- Chocolate cake mix contains sugar.
- Sugar has oxygen, hydrogen, and carbon in it.

Next, help the students think specifically about what their data show. This is an important critical thinking step that will help them evaluate future experiments.

III. Conclusions

Have the students review the results they recorded for the experiment. Have them draw conclusions based on the data they collected.

Help them write concluding statements that are valid. Encourage them to avoid stating opinions or any conclusions that cannot be drawn strictly from their data.

For example, it may be true that all cereals contain salt. However, this particular investigation cannot confirm or deny that conclusion. The most that can be stated from this investigation is "Brand X contains salt and Brand Y contains salt," but any further statement is conjecture.

Help them formulate their conclusions using the words some, all, many, and none. Point out that the statement, "All cereals contain salt," is not valid, but based on this investigation, it is valid to say, "Some cereals contain salt."

Again, there are numerous possible answers. One student may list "sugar" as a component in soup, and another may list "salt," and both answers could be correct. The true test is whether the statements about the data are valid or not valid.

Also, try to show students where broad statements can be made validly. For example, "All recent U.S. pennies contain copper" is probably a valid statement even though we haven't checked every U.S. penny.

This may seem fairly subtle, but the main point is to help them understand the kinds of valid conclusions science can offer based on scientific investigation.

IV. Why?

Read this section of the *Laboratory Notebook* with your students.
Discuss any questions that might come up.

V. Just For Fun

Students are to select one item from their list and do research to find out as much as they can about how it was made, where it was made, and where the different components might have come from.

Experiment 4

Modeling Molecules

Materials Needed

- small, colored marshmallows, 1 pkg
- large marshmallows, 1 pkg (could also use gumdrops and/or jellybeans in place of marshmallows)
- toothpicks, 1 box

Optional

- food coloring

Objectives

In this experiment students will explore how atoms combine to make molecules.

The objectives of this lesson are for students to:

- Observe how making models is helpful in understanding how atoms combine to form molecules.
- Explore the concept that atoms follow rules when combining to make molecules.

Experiment

I. Think About It

Read this section of the *Laboratory Notebook* with your students.

Ask questions such as the following to guide open inquiry.

- *Do you think it's important for people to have rules to follow? Why or why not?*

- *What do you think life would be like if people did not have any rules?*

- *Do you think atoms have rules to follow when they combine to make molecules? Why or why not?*

- *What do you think life on Earth would be like if atoms could combine in any way they wanted to with no rules?*

II. Experiment 4: Modeling Molecules

In this experiment students will use marshmallows and toothpicks to explore the ways in which atoms combine to form molecules.

Have the students read the entire experiment.

Objective: An objective is provided.
Hypothesis: Have the students write a hypothesis. Some examples:

- *Models can show how atoms combine.*

- *By making models, I can see how atoms follow rules.*

EXPERIMENT

❶-❷ Two sizes of marshmallows are preferred for this experiment. Gumdrops and jellybeans can also be used.

Have the students make marshmallow and toothpick molecules without following any rules. Encourage the students to make molecules of various sizes and shapes. They do not need to record every shape they make, but have them draw at least several different shapes.

All of their answers will be correct since all shapes are valid in this step. Encourage them to use their imagination in combining the marshmallows.

❸   Students will make "real" molecule models following specific rules.

The rules for carbon, nitrogen, oxygen, hydrogen, and chlorine are shown. Note that the orientation of the bonds (toothpicks) is also important. Before making molecules, the students can first practice putting the toothpicks into several marshmallows while following these rules.

Note that the large marshmallows are assigned to the atoms carbon, nitrogen and oxygen. If this is confusing, try to differentiate between the atoms by color-coding each with a drop of food coloring.

Results

❶   Now the students will follow the rules and make "molecules" with the marshmallow "atoms." Have them draw their molecules.

These illustrations show the correct shapes for the molecules that are given. (Drawings may vary.)

Have the students note the number of bonds for each molecule, and ask them whether or not they followed the rules.

(Drawings may vary.)

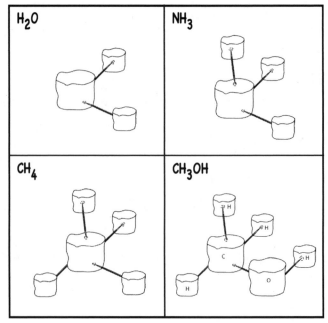

III. Conclusions

Have the students write some conclusions about the molecules they have created with the marshmallows. Help them try to be specific with the conclusions they write.

IV. Why?

Read this section of the *Laboratory Notebook* with your students.
Discuss any questions that might come up.

V. Just For Fun

Students will follow the rules and make their own "molecules." For each molecule model, have the students note how many bonds each "atom" forms and how many of each type of atom are in the molecule.

Have them draw their molecule models.

Some suggested molecules are the following:

CCl₄ four chlorine atoms attached to one central carbon (carbon tetrachloride)

CH₃CH₃ two carbon atoms connected to each other with three hydrogens each (ethane)

CH₃CH₂OH two carbon atoms connected to each other. One carbon atom has three hydrogens attached. The other has two hydrogens and an oxygen attached to it, and the oxygen has a hydrogen attached. (ethanol)

The students can build many different molecules while still following the rules.

(Answers will vary. Some examples are shown.)

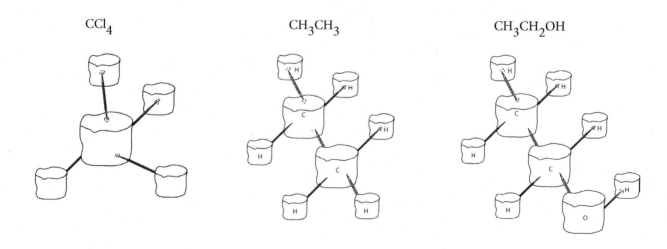

CCl₄   CH₃CH₃   CH₃CH₂OH

Experiment 5

Identifying Chemical Reactions

Materials Needed

- baking soda
- lemon juice
- balsamic vinegar
- salt and water:
  15–30 ml salt dissolved in
  120 ml water
  (1–2 tbsp. salt dissolved in
  1/2 cup of water)
- 2 or more egg whites
- milk
- small jars, 7 or more
- measuring cups and spoons
- eye dropper

Peanut Brittle

- 360 ml (1 1/2 cups) sugar
- 240 ml (1 cup) white corn syrup
- 120 ml (1/2 cup) water
- 360 ml (1 1/2 cups) raw peanuts (can be omitted)
- 5 ml (1 teaspoon) baking soda
- buttered pan

Objectives

In this experiment students will observe how chemical reactions cause bonds to form and break, creating changes that can be seen.

The objectives of this lesson are for students to:

- Observe evidence that chemical reactions are taking place.
- Chart their observations.

Experiment

I. Think About It

Read this section of the *Laboratory Notebook* with your students.

Ask questions such as the following to guide open inquiry.

- *Do you think chemical reactions are important in your daily life? Why or why not?*

- *Do you think any two liquids mixed together will undergo a chemical reaction? Why or why not?*

- *If you mix two liquids together and they have a chemical reaction, do you think there will be a chemical reaction every time more of those liquids are mixed? Why or why not?*

- *Do you think you can tell whether or not a chemical reaction has happened? Why or why not?*

- *What clues would you look for to show that a chemical reaction has taken place?*

- *What clues would you look for to show that a chemical reaction has not taken place?*

II. Experiment 5: Identifying Chemical Reactions

In this experiment students will examine chemical reactions and try to identify when they occur. Balsamic vinegar is recommended because the reaction when it is mixed with baking soda will be more dramatic, but other kinds of vinegar may be used.

Have the students read the entire experiment.

Objective: An objective has been provided.
Hypothesis: A hypothesis has been provided

EXPERIMENT

❶ Have the students put a small amount of each substance listed into its own jar, and then have them examine the contents of each jar, taking note of the properties of each substance. Have them record the color, texture, and odor next to each item on the materials list. For example:

- Baking Soda: white powder, no odor.
- Balsamic Vinegar: dark liquid, sour odor.

Although most of the items are food items, **do not allow the students to taste them** since tasting is not part of this experiment.

A variety of food items in addition to those on the materials list may be used. Bleach and ammonia cause good chemical reactions, but they can give off strong odors and so these chemicals are NOT RECOMMENDED.

❷ Have the students write the "reagents" (chemicals used in chemical reactions) on the top and side of the grid provided.

Have them mix some of each of two substances together in a clean jar and observe the results. Have them rinse the mixing jar in between tests. It might be interesting to the students to check the order of addition—e.g., add lemon juice to baking soda, then add baking soda to lemon juice—to see if a difference can be observed (the order of addition should not matter). This is optional.

❸ Have the students record their observations in the appropriate box for each reaction.

Results

(Expected results)

	milk	lemon juice	salt water	baking soda	balsamic vinegar	egg whites
milk		REACT precipitate	NO	NO	REACT precipitate	NO
lemon juice			NO	REACT precipitate	NO	REACT precipitate
salt water				NO	NO	NO
baking soda					REACT precipitate	NO
balsamic vinegar						REACT precipitate
egg whites						

Results for Unknown Solutions

❶ Give the students two "unknown" solutions—ones that were used in the experiment but that you are not now identifying. They can either be two substances that will react or two that won't react. Have students describe the properties of each unknown.

This part of the experiment can be done more than once. In addition, you can have the students give you "unknowns" to see if you can identify them.

The students have observed all of the reactants both before and after a reaction. They now have the necessary knowledge to identify an unknown.

❷ Have the students mix the two substances and describe the results, including why they think a chemical reaction did or did not take place.

❸ Have the students record what they think the two substances are and how they identified them.

An option for additional experimentation is to give the students only one unknown. Have them guess what it might be before performing any tests. Then have the students test this unknown with each of the other reactants. Have them prove the identity of the unknown with the chemical reactions they have already observed.

III. Conclusions

Have the students review the results they recorded for the experiment and write valid conclusions. Help them state conclusions that reflect only the data found in this experiment. For example, "Salt water does not react with anything" is not a valid conclusion because we haven't tested all substances. However, "Salt water does not react with any of the items we tested" is valid.

IV. Why?

Read this section of the *Laboratory Notebook* with your students.
Discuss any questions that might come up.

V. Just For Fun

Help the students make peanut brittle following the directions provided. They will be able to observe baking soda decompose (undergo a decomposition reaction), giving off carbon dioxide gas while the peanut brittle is being made.

CHEMISTRY

Peanut Brittle

360 ml (1 1/2 cups) sugar

240 ml (1 cup) white corn syrup

120 ml (1/2 cup) water

360 ml (1 1/2 cups) raw peanuts

5 ml (1 teaspoon) baking soda

buttered pan

Boil sugar, water, and syrup in a sauce pan over medium heat until the mixture turns a little brown. Add 360 ml (1 1/2 cups) raw peanuts Stir until golden brown. Don't over-brown. Add 5 ml (1 teaspoon) baking soda. Spread on buttered pan.

Have students make observations about the evidence of a chemical reaction occurring. Have them think about what other chemical reactions they have noticed when food items are being cooked or after they are cooked.

Have them record their observations and ideas in the space provided. There are no "right" answers.

Experiment 6

Putting Things in Order

Materials Needed

- pencil and eraser

Objects chosen by students, such as:

- rubber ball
- cotton ball
- orange
- banana
- apple
- paper
- sticks
- leaves
- rocks
- grass
- Legos
- building blocks
- other objects

Optional

- several sheets of paper

Objectives

In this experiment students explore categorizing objects by their features.

The objectives of this lesson are for students to:

- Explore how objects can be categorized in different ways and chart their data.
- Observe the difficulties of categorizing objects.

Experiment

I. Think About It

Read this section of the *Laboratory Notebook* with your students.

Ask questions such as the following to guide open inquiry.

- *What are some groups of objects you can think of?*
- *How would you decide which objects should go in each group?*
- *Do you think it can be helpful to you put objects into groups? Why or why not?*
- *Do you think some objects can go into more than one group? Why or why not?*
- *Do you think it is easy or difficult to put objects in groups? Why?*
- *How do you use groups in your day-to-day life?*

II. Experiment 6: Putting Things in Order

In this experiment, students will try to organize different objects according to their characteristics, such as shape, color, or texture. There are no "right" answers for this experiment, and the categories the students choose will vary.

Have the students read the entire experiment.

Help them collect a wide variety of objects of their choice that they will categorize.

BIOLOGY

Objective: Have the students write an objective. Some examples:

- *To put objects into different categories.*

- *To use categories and subcategories.*

Hypothesis: Have the students write a hypothesis. Some examples:

- *It will be easy to put objects in categories.*

- *Some objects will go into more than one category.*

EXPERIMENT

❶ Have the students place the collected objects on a table and then make careful observations. Guide them to notice some features of the objects, such as color, shape, and texture. Also, discuss any common uses, for example, those used as toys or those used as writing instruments.

❷ Have the students fill in the chart provided, listing each object and a few of its characteristics. Help them to be as descriptive as possible. For example, oranges can be described as round, orange, sweet, food, living, etc. Tennis balls are round, fuzzy, yellow or green (or another color). It is not necessary for them to fill in all the lines on the chart.

❸ Next, have the students determine some overall categories into which the objects can be placed. For example, marbles, cotton balls, and oranges are round, so "Round" could be a category. Basketballs, baseballs, and footballs are all balls, so another category could be "Types of Balls." Have the students write a category at the top of each column using a PENCIL so they are able to change the categories as more items are being written down.

❹ Students will list objects in the category that describes them according to their characteristics. Some items may fit into more than one category. Basketballs can fit into both the category "Round" and the category "Types of Balls." In the chart provided, have the students write down each item in all of the categories where it fits.

❺ Have the students look at each category separately and then choose three categories to further divide into subcategories. Guide them in thinking about what the subcategories might be, trying to choose categories that allow all of the items to ultimately be listed. If necessary, they can rename some of the main categories to better fit the items listed. The names of the categories and subcategories can be adjusted as needed so that each item is listed in a category and subcategory, but it's possible that not all of the items can be placed in a category and a subcategory. This can be quite challenging. The point of this exercise is to illustrate the difficulty of trying to find a suitable organizational scheme for things with different characteristics.

III. Conclusions

Have the students review the results they recorded for the experiment. Help them write valid conclusions based on the data they have collected. For example:

- Both oranges and cotton balls are round.
- Both cotton balls and marshmallows are white.
- Tennis balls and cotton balls are both fuzzy.

Examples of conclusions that are not valid:

- Both cotton balls and marshmallows are white. Marshmallows are sweet so cotton balls are sweet.
- Tennis balls and cotton balls are both fuzzy. Tennis balls are bouncy so cotton balls must be bouncy.

It is important to use only the data that has been collected and not make statements about the items that are not backed up by the data. It is obvious that marshmallows and cotton balls are both white, but it is not true that cotton balls are sweet. Because two or more items have one or two things in common does not mean that all things are common between them. Discuss this observation with the students.

Discuss the difference between valid and invalid conclusions. A valid conclusion is a statement that generalizes the results of the experiment, but draws only from the data collected. It does not go beyond the results of the data to include things that haven't been observed and does not connect results that should not be connected. An invalid conclusion is a statement that has not been proven by the data, or a statement that connects the data in ways that are not valid. The example given is that marshmallows are sweet and white, but although cotton balls are also white, it is invalid to say they are sweet like marshmallows.

IV. Why?

Read this section of the *Laboratory Notebook* with your students.
Discuss any questions that might come up.

V. Just For Fun

Students are to list 15 or more living things that can be seen without a microscope or magnifying glass. Then they will create their own taxonomic system to categorize them. There are no "right" answers.

Have them record their chart. They may want to use more paper.

What's in Spit?

Materials Needed

- tincture of iodine [**VERY POISONOUS—DO NOT ALLOW STUDENT TO EAT** any food items that have iodine on them]
- bread, 1 slice
- timer
- wax paper
- marking pen
- cup
- refrigerator
- a green vegetable
- one or more other vegetables or fruits

Objectives

In this experiment students will observe a chemical reaction in part of the metabolic process.

The objectives of this lesson are for students to:

- Observe evidence of a chemical reactions that happens in the body.
- Observe how the process of digestion of food begins.

Experiment

I. Think About It

Read this section of the *Laboratory Notebook* with your students.

Ask questions such as the following to guide open inquiry.

- *How do you think your body digests food?*

- *Why do you think food needs to be chewed?*

- *Why do you think you have saliva in your mouth?*

- *Do you think your body could digest food if you didn't have saliva? Why or why not?*

- *Do you think digestion requires chemical reactions? Why or why not?*

- *Do you think chemical reactions happen in your mouth? Why or why not?*

II. Experiment 7: What's in Spit?

In this experiment students will investigate the part of the digestive process carried out by proteins in saliva. Have the students read the experiment before writing an objective and hypothesis.

EXPERIMENT

An example **Objective:** *We will investigate what saliva does to bread.*
An example **Hypothesis:** *We will be able to test changes in the bread by using iodine.*

❶ Have the students break the bread into several small (bite size) pieces.

❷ Students will chew one piece of bread for 30 seconds, another piece for 1 minute, and a third for several minutes. Have them set a timer for each.

❸ After each chewing time is up, have the students spit the chewed bread onto a piece of wax paper and use a marker to record the length of time it was chewed.

❹ Have the students place one small piece of unchewed bread next to each piece of chewed bread.

❺ Have the students put a drop of iodine on each of the pieces of bread, both chewed and unchewed.

❻ Have them record their results in the chart provided. They should observe that the color resulting from the iodine reacting with the bread that has been chewed for longer times is not as black as with unchewed bread or bread that has not been chewed as much.

❼ Have the students collect saliva by spitting into a cup several times. Then they will take two small pieces of bread and soak both in the saliva. They can add more saliva to the cup if needed.

❽ Have the students place each piece of soaked bread on a separate piece of wax paper and put one in the refrigerator and leave one out at room temperature.

❾ After 30 minutes, have the students test each piece of bread by putting a drop of iodine on each.

❿ Have them record their results. The refrigerated bread should turn more black than the unrefrigerated bread because the cold temperature slows the chemical reaction.

III. Conclusions

Have the students review the results they recorded for the experiment. Have them draw conclusions based on the data they collected.

IV. Why?

Read this section of the *Laboratory Notebook* with your students.
Discuss any questions that might come up.

V. Just For Fun

Have the students repeat the experiment with celery, kale, or another green vegetable. The iodine should not change color. Have them test one or more other vegetables or fruits. Following are some expected results, but any vegetables or fruits can be tested:

Color change: apple, banana, pasta, potato, yam

No color change: celery, kale, spinach, green bell pepper

Have the students draw conclusions from their results.

Experiment 8

Inside the Cell

Materials Needed

- pencil
- colored pencils or crayons
- student-selected materials to build a model cell

Objectives

In this experiment students will explore cells and their structure.

The objectives of this lesson are for students to:

- Observe that cells are highly complex and highly ordered.
- Compare the features of three different types of cells, observing similarities and differences.

Experiment

I. Think About It

Read this section of the *Laboratory Notebook* with your students.

Ask questions such as the following to guide open inquiry.

- *Why do you think scientists study cells?*
- *Do you think it is helpful for scientists to put different kinds of cells into groups? Why or why not?*
- *Do you think it's important that there are different kinds of cells? Why or why not?*
- *What do you think life would be like if all cells were exactly the same?*
- *How do you think you could tell one type of cell from another?*

II. Experiment 8: Inside the Cell

In this exercise students will examine the similarities and differences between three cell types.

All cells share some common features. One such feature is DNA (deoxyribonucleic acid). DNA is often referred to as the genetic code. Almost every cell has DNA. The DNA in a cell contains many volumes of information that the cell needs to make proteins, metabolize nutrients, grow, and divide.

Another feature common to cells is that they have ribosomes which make proteins from RNA (ribonucleic acid). RNA is different from DNA but is still a nucleic acid. RNA is made from DNA and proteins are made from RNA.

DNA->RNA->proteins

In living cells there are no known exceptions to this paradigm. Proteins are always made from RNA, and the RNA used to make proteins is always made from DNA.

Have the students read the entire experiment.

Objective: An objective is provided.

EXPERIMENT

The first part of the experiment has questions for students to answer. Following are examples of possible answers. Answers may vary.

- *List some things you observe in the drawings in the textbook that are similar for all three cell types:*

  (Examples. Answers may vary.)
  All cells contain DNA.
  All cells contain ribosomes.
  All cells have something that holds them together, like a cell wall or plasma membrane.

- *List some observations of things that are different:*

  (Examples. Answers may vary.)
  Bacterial prokaryotic cells do not have a nucleus.
  Animal cells do not have a cell wall.
  Plant cells contain chloroplasts, but animal cells do not.

- *List the function of each of the following:*

  nucleus   In eukaryotic cells, the nucleus holds together the DNA and the proteins needed to use the DNA.

  mitochondria   organelles that make energy; found in plant and animal cells

  chloroplasts   organelles that use the Sun's energy to make food; found in plant cells

  cell wall   stiff outer membrane found in plant cells that makes the plant sturdy

  lysosome   the place where big molecules get broken down

  peroxisome   the place where poisons in the cell are removed

❶ Students are asked to list differences between bacteria, plants, and animals and why their cells may need to be different. There is a chart provided to be filled in. Answers will vary and there are no "right" answers.

To learn more, students can do research online or at the library.

Following are some facts about bacteria:

- They can be spherical, rod shaped, or spiral.
- They live in many different environments including soil, water, organic matter, and in the bodies of plants and animals.
- They make their own food (are autotrophic), or live on decaying matter (are saprophytic), or live off a live host (are parasitic).
- They can be either beneficial or harmful to humans.

Some facts about plants and animals:

Plants have organs, as do animals, and therefore need to be made of many different types of cells. Have the students think about the different parts of a plant, such as the leaves and roots, and discuss what the cells of each part might need to do. (For example: Root cells need to take up minerals from the soil. Since roots are in the dirt, they do not have to be green like leaves. The leaves are green because they need to use chloroplasts for collecting light.)

Discuss how plant cells differ from animal cells. For example, plants don't have bones, and they don't usually move, so they don't need muscles like some animals. Have the students think of a variety of animals, such as deer, fish, and frogs, and then discuss the differences between them. Next, have them write down why there are different types of cells in these different creatures.

❷ Bacterial prokaryotic cell ❸ Animal cell ❹ Plant cell

Have the students fill in the blanks in the drawings on these pages. Have them first identify what type of cell they are looking at. Have them label as much of the drawings as they can without looking at the *Student Textbook* and then refer to the textbook to finish the labeling. When they have filled in all the labels, have them color the different parts of the cell. The colors do not need to match those in the textbook.

As the students fill in the blanks, discuss the functions of the various parts of the cell. Point out how the structures differ and, where possible, point out how the structure of the part matches its function. For example, the flagellum looks like a whip and is used in a whipping motion for swimming. The cell membrane and cell wall are used to enclose the contents of the cell, are thin, and extend around the outside of the cell.

III. Conclusions

Have the students arrive at some conclusions about cells based on what they have learned in this chapter.

Some examples include:

- *All living things are made of cells.*
- *All living things have DNA.*
- *Not all cells are alike.*

Answers may vary.

IV. Why?

Read this section of the *Laboratory Notebook* with your students.
Discuss any questions that might come up.

V. Just For Fun

Students will make a simple model of a cell. They can select a few features to model rather than trying to include them all.

Have them decide which of the three cells they would like to model and then make a list of materials they think they could use to build their model. They can use one type of material, such as colored clay, or a combination of materials, such as clay, paperclips, string, wire, food items, etc. Encourage them to use their imagination in selecting materials to represent the different cell features.

BIOLOGY

Wash Your Hands!

Materials Needed

- dehydrated agar powder*
- distilled water
- cooking pot
- measuring spoons
- measuring cup
- cup
- plastic petri dishes (20)**
- cotton swabs
- permanent marker
- oven mitt or pot holder

* http://www.hometrainingtools.com/
nutrient-agar-8-g-dehydrated/p/
CH-AGARN08/

** A stack of 20 can be ordered from:
http://www.hometrainingtools.com/petri-
dishes-plastic-20-pk/p/BE-PETRI20/

Objectives

In this experiment students will grow bacterial cultures and practice using controls.

The objectives of this lesson are for the students to:

- Explore how the use of control experiments is required to verify data.
- Observe how scientific explanations emphasize evidence and use scientific principles.

Experiment

I. Think About It

Read this section of the *Laboratory Notebook* with your students.

Discuss how bacteria and viruses are small organisms that we can't see with our eyes alone and that they can live on surfaces and inside the body.

Discuss how some bacteria are healthful and help us digest our food and some bacteria are harmful and make us sick.

Explain that scientists use agar plates to grow bacteria. Agar is made from various types of seaweed and is used as a thickening agent. The dehydrated agar used in this experiment is called nutrient agar and contains not only agar but also vitamins, amino acids, carbon, and nitrogen derived from beef extract. Nutrient agar will give off a slight odor when cooked.

Ask questions such as the following to guide open inquiry.

- *Do you think bacteria are important? Why or why not?*
- *Do you think all bacteria are harmful? Why or why not?*
- *Do you think you can ever see bacteria? Why or why not?*
- *Why do you need to wash your hands after being outside?*
- *Why do you need to wash your hands after using the bathroom?*
- *Do you think your hands are "clean" after you wash them? Why or why not?*
- *Do you think the surfaces in your house, like the computer terminal and kitchen doorknob might have bacteria on them? Why or why not?*

II. Experiment 9: Wash Your Hands!

Have the students read the entire experiment.

Objective: Have the students write an objective.
Hypothesis: Have the students write a hypothesis.

EXPERIMENT

Part I: Preparing Agar Plates

Help the students assemble the materials for making agar plates. Making agar plates is not difficult but may take a little practice. Keep the experimental area as clean as possible to avoid contamination.

❶ Have the students prepare a clean, flat surface on which to pour the plates. Have them spread out 18-20 petri dishes to use for both the *Wash Your Hands* experiment and the *Just For Fun* experiment.

❷ Have the students follow the directions for preparing the agar.

❸ The hot agar can be cooled slightly before pouring but should not cool so much that it starts to harden. Have the students use an oven mitt or pot holder while picking up the pot of agar. To prevent contamination, have them slide the lid partially off a petri dish just before they fill it and then re-cover it immediately after it is filled. Have them carefully pour a small amount of the hot agar into each of the petri dishes. It is easy to pour too much agar in one petri dish and too little in another, but they should try to put just enough agar in each petri dish to cover the bottom.

❹-❻ Once the agar has solidified, the petri dishes can be stored upside down with the agar on top. This keeps condensation from collecting on the surface of the agar. Have the students stack the inverted petri dishes and place them in the refrigerator.

Part II: Testing for Bacteria

❶-❷ The students will prepare two petri dishes to act as controls for the experiment. The control plates will indicate if their agar or their water is contaminated. If the agar control shows growth after a few days, this means the agar was contaminated during preparation. The experiment can be repeated. Be careful to use a clean area. If the water control shows growth after a few days, the water can be boiled and the experiment repeated. Most of the time, both the agar and the water controls will be clean. Bacterial growth occurring after several weeks is normal.

Have the students take one petri dish from the refrigerator, label it "Agar," and return it to the refrigerator. A second petri dish will be labeled 'Water" and left out at room temperature.

❸ Students will prepare the control test for the water being used. Have them swirl a cotton swab in the distilled water and shake off the excess water. Then have them "streak" the plate labeled "Water" by gently moving the swab in a zigzag motion across the agar from one side of the petri dish to the other, being careful not to break the hardened surface of the agar.

❹-❺ They will now test their hands for the presence of bacteria. Have them remove another petri dish from the refrigerator and label it "Hands." Have them take an unused cotton swab, swirl it in the distilled water, and shake off the excess Then they will rub the swab on their fingertips and streak the agar plate as in Step ❸.

❻ Have the students choose some surfaces they'd like to test and write the names of the surfaces in the chart provided.

❼ Have the students remove the petri dishes one at a time from the refrigerator as they test each surface they've chosen. The plate labeled "Agar" should be left in the refrigerator. Have the students refer to their chart, and keeping the petri dish agar side up, label it with the name of the surface to be tested. The streaked plates will be left out at room temperature.

❽ Have the students prepare each plate using the same method as before.

❾ Once all the plates have been prepared, have the students put them in a stack agar side up. Have them remove the "Agar" plate from the refrigerator and add it to the stack. The plates will be left to incubate for a week to ten days. They should not be put in a bag.

Results

After 7-10 days, have the students examine each of their plates and record their observations in the table provided.

III. Conclusions

Have the students review the results they recorded for the experiment. Have them note whether there was bacterial growth on the control plates. Then have them compare their control plates to their test plates to look for bacterial growth. Have them draw conclusions based on their data—what they actually observed and not what they think should have happened.

IV. Why?

Read this section of the *Laboratory Notebook* with your students.
Discuss any questions that might come up.

V. Just For Fun

Have the students retest the surfaces that showed bacterial growth in Part II of the experiment. This time they will wash each surface with a different household cleaner and repeat the experiment to see if they can determine how effective the different cleaners are at removing bacteria. A table is provided for them to record their results.

Experiment 10

It's the Law!

Materials Needed

- tennis ball
- paperclip
- yarn or string (about 3 meters [10 ft])
- marble
- bouncing ball, 1 (or 2 or more of different sizes)

Optional

- penknife, ice pick, awl, or other sharp tool
- pliers

Objectives

In this experiment students will be introduced to the concept of *physical laws*—a fundamental concept in physics.

The objectives of this lesson are for students to:

- Use the scientific method to observe the physical world.
- Explore Newton's First Law of Motion.

Experiment

I. Think About It

Read this section of the *Laboratory Notebook* with your students.

Ask the students what a law is, such as a law against driving too fast or a law against stealing. Ask if these laws can be broken and, if so, why they can be broken.

Explain that laws in physics differ from the kinds of laws that govern our country. In physics a law is an overall principle or relationship that remains the same and is not broken.

Ask the students to describe several observations they have made about how objects behave in the physical world. Encourage them to discuss as many observations as they can think of. There are no "right" answers, and at this point, it is not important to know the reasons why something happens.

Ask questions such as the following:

- *What happens when you put on the brakes while riding a bicycle? Do the tires stop immediately? Do they skid?*

- *What happens when you throw a ball into the air? Does it reach the clouds? Does it come down in the same spot?*

- *What happens when you turn on a flashlight? How far can you see the light? Can you see the beam from a flashlight in the daytime?*

- *Have you ever thrown a ball and had it not come down (except when it gets stuck somewhere like in a tree)?*

- *Does ice always float?*

- *Does the Sun always come up in the morning?*

PHYSICS

II. Experiment 10: It's the Law!

Read this section of the *Laboratory Notebook* with your students.

In this experiment students will discover Newton's First Law of Motion by observing the movements of a tennis ball and a marble. Newton's First Law of Motion can be stated as:

> *A body will remain at rest or in motion until it is acted on by an outside force.*

The objective is provided. Have the students read through the experiment and then write a hypothesis based on the steps of the experiment.

Part I

❶ Students are to observe how a ball travels through the air. They should notice that the ball will go up and come down in some kind of arc every time they throw it. The arc can be shallow or sharp depending on how they throw the ball.

Challenge them to throw the ball so that it won't come down.

Ask them if they can get the ball to go up and down in a pattern different from an arc.

❷ Have the students follow the directions to use a paperclip to attach the string to the tennis ball. It is somewhat difficult to puncture the tennis ball with the paperclip, so have students take care while doing this. You may want to first put a small hole in the tennis ball with a penknife, ice pick, or awl before having the students insert the paperclip.

Alternatively, a longer string can be used, wrapped several times around the ball, and secured with tape. It is harder to get the string to stay attached to the ball using this method.

❸ With the string attached, the trajectory of the tennis ball will be different. When the string has reached its full length, the ball will abruptly stop and fall to the ground.

Have the students throw the ball several times. Ask them if they can change how the ball falls to the ground. They should notice that if they shorten the string, the ball does not travel as far as when the string is longer. They should also notice that if they do not throw the ball very far and it does not reach the end of the string, the ball will travel almost as if there were no string attached to it.

Have them record their results.

Part II

❶ Students will roll a marble several times on a smooth surface and record their results.

❷ Students will roll a marble several times on a rough surface and record their results.

III. Conclusions

Have the students review the results they recorded for Part I and Part II of the experiment. Have them draw conclusions based on the data they collected.

IV. Why?

Read this section of the *Laboratory Notebook* with your students.
Discuss any questions that might come up.

V. Just For Fun

In this experiment students will play with a bouncing ball and observe how the amount of force used changes the way the ball bounces. With more force, the ball will bounce higher and more times.

If bouncing balls of different sizes are available, have the students repeat the experiment and observe whether the size of the ball affects the outcome.

PHYSICS

Experiment 11

Fruit Works?

Materials Needed

- Slinky
- several paperclips
- 1-2 apples
- 1-2 lemons or limes
- 1-2 oranges
- 1-2 bananas
- spring balance scale or food scale
- meterstick, yardstick, or tape measure
- tape

Objectives

In this experiment students will be introduced to the fundamental concepts of force, energy, and work.

The objectives of this lesson are for students to:

- Gain a basic understanding of the concepts of force, energy, and work.
- Observe gravitational force acting on an object.

Experiment

I. Think About It

Read this section of the *Laboratory Notebook* with your students.

Have a discussion with the students concerning their own ideas about force, energy, and work. Ask questions such as the following to guide open inquiry.

- *What do you think work is?*
- *Do you think work is done if you move bricks from the front yard to the back yard? Why or why not?*
- *Do you think work is done if you lift a book? Why or why not?*
- *Do you think more work is done if you lift three books at the same time than if you lift one book? Why or why not?*
- *Is work done if a piece of fruit is dropped? Why or why not?*
- *What do you think force is?*
- *Can you give some examples of force?*

II. Experiment 11: Fruit Works?

Read this section of the *Laboratory Notebook* with your students.

In this experiment the students will try to determine how much work a variety of fruits can do. Remind the students that:

work = distance x force

Objective: Have the students read the entire experiment and then guide them to think of a possible objective. For example:

> • *Using a Slinky, we will find out if a banana can do more work than an orange.*
>
> • *We will measure the work that fruit can do.*
>
> • *We will find out if two bananas do more work than one.*

Experiment

❶ Have the students "weigh" the pieces of fruit by picking them up. They should be able to tell which one is the heaviest and which is lightest just by feeling the weights of the different fruits in their hands. Have the students make a guess about which fruit will do more work and which will do the least.

❷ Have them state their theory as the hypothesis. For example:

> • *A banana is heavier than a lemon and will do more work.*
>
> • *The orange is lighter than the apple and will do less work.*
>
> • *The apple and the orange weigh the same amount and will do the same amount of work.*

❸-❹ Have the students use a scale to weigh each piece of fruit and record the weight in the chart provided.

❺ Have the students make hooks from paperclips to attach the fruit to the Slinky. We found that the paperclips worked fairly well, but younger kids found tape easier to use. The fruit can be fixed to the Slinky in any manner. You might ask the students to come up with their own ideas for attaching the fruit.

❻ Students will have to experiment with the Slinky and the number of coils that hang down. We found that it worked fairly well to have a student hold most of the coils in their hand and allow only a few coils to fall below the hand. Also, instead of holding the Slinky, it can be attached to a branch of a tree or a fixed ledge of some sort. Just make sure the Slinky is free to extend and does not contact any other surface and that the Slinky is at the same distance from the floor each time.

❼ Have the students measure the distance from the floor to the bottom of the Slinky and record this number in the chart provided.

PHYSICS

❽ Have the students attach a piece of fruit to the last coil of the Slinky, and allow the coils to extend.

❾ Have them measure the distance from the ground to the bottom of the Slinky for each piece of fruit that is tested and then record the distance in the chart provided.

❿ Have the students repeat Steps ❽ and ❾ with different kinds of fruit.

Results

❶ Have the students use the chart in Step ❾ of the previous section. They will subtract the distance from the floor to the Slinky without any fruit on it from the distance with the fruit on it. The result will be the distance the Slinky extended.

❷ Students will use the following equation to determine how much work was done by each piece of fruit: *work = distance x force* where force is the weight of the fruit.

Explain the measurement system the students are using:

When using the metric measurement system in the equation above, the unit of measure of work is the kilogram-meter. For example, 2 kilograms x 2 meters = 4 kilogram-meters of work. (A gram is equivalent to .001 kg.)

When using the British measurement system in the equation above, the unit of measure of work is the foot-pound. For example, 2 pounds x 2 feet = 4 foot-pounds of work. (An ounce is equivalent to .0625 lb.)

NOTE:

In this experiment weight and mass are being used interchangeably, even though they are not the same thing. Mass is the property that causes objects to have inertia. Mass and inertia will be discussed in a later book.

Technically, when we weigh something we are measuring its mass times the force of gravity (gravitational acceleration). Gravitational acceleration is equal to 1 and is the same everywhere on Earth. For this reason and for the purposes of this experiment we therefore use mass and weight interchangeably.

III. Conclusions

Have the students review the results they recorded for the experiment. Have them draw conclusions based on the data they collected.

IV. Why?

Read this section of the *Laboratory Notebook* with your students.
Discuss any questions that might come up.

PHYSICS

V. Just For Fun

❶-❷ Have the students predict what would happen if they attached two bananas of about the same size or two other pieces of the same kind and similar size of fruit to the Slinky. They should predict that two pieces of fruit will do more work than one piece of fruit. Have them test this prediction by attaching two pieces of fruit to the Slinky and repeating the steps done previously. Have them record their results and then calculate the amount of work done. Have them analyze their results. Do two pieces of fruit do twice the work? Three times the work? Four times the work?

Experiment 12

Smashed Banana

Materials Needed

- small to medium size toy car
- stiff cardboard
- wooden board, smooth and straight (more than 1 meter [3 feet] long)
- straight pin or tack, several
- small scale or balance
- one banana, sliced
- 10 pennies
- meterstick, yardstick or tape measure
- tape

Objectives

Performing this experiment will help students understand that energy exists in different forms, is converted from one form to another, and can do work.

The objectives of this lesson are for students to:

- Observe energy changing from one form to another.
- Use a formula to calculate gravitational potential energy.

Experiment

I. Think About It

Read this section of the *Laboratory Notebook* with your students.

Ask questions such as the following to guide open inquiry.

> - *What do you think it would mean if someone said to you that you have the potential to become a famous scientist?*
>
> - *What other examples of potential can you think of?*
>
> - *What do you think gravitational potential energy is?*
>
> - *Do you think gravitational potential energy can do work? Why or why not?*
>
> - *What do you think kinetic energy is?*
>
> - *Do you think gravitational potential energy and kinetic energy are the same or different? Why?*

II. Experiment 12: Smashed Banana

Read this section of the *Laboratory Notebook* with your students.

❶ Have the students read the entire experiment.

Objective: Have the students write an objective. Some examples:

> - *We will measure how much GPE is needed to smash a banana.*
>
> - *We will show that a heavier toy car needs less height (less GPE) to smash a banana.*

PHYSICS

Hypothesis: Have the students write a hypothesis. For example:

> - *The toy car will not be able to smash the banana no matter how high the ramp.*
>
> - *The toy car will smash the banana when the ramp is two or three feet high.*
>
> - *The toy car is not big enough to smash the banana.*
>
> - *The toy car needs to have at least 50 pennies on it to smash the banana.*

❷-❸ Have the students assemble the apparatus according to the directions in the *Laboratory Notebook*. The experiment works best if the board is smooth and reasonably straight. A cardboard tube, such as a wrapping paper tube, also works if cut lengthwise and opened up to make a trough. The car should have good wheels and roll smoothly and easily to reduce friction.

When testing this experiment, we found that we needed to put several pieces of banana next to each other at the bottom of the ramp since the car often does not travel in a straight line.

❹   Have the students weigh the toy car and record the weight in the space provided.

❺-❼ Have the students elevate one end of the ramp 5 cm (2 inches) and roll the toy car down the ramp. Then have them elevate the ramp in 5 cm (2 inch) increments, allowing the car to travel down the ramp and hit the banana each time the ramp is raised. Have them record the results each time.

We found that an average toy car does not really smash the banana until the ramp has been elevated more than 30 cm (12 inches).

❽   Have the students answer the questions.

❾   The students will add pennies to the car to make it heavier. They may need to tape the pennies to the car. Have them weigh the car again with the pennies on it and repeat the experiment. They should discover that the ramp will not need to be elevated quite as high in order for the toy car to smash the banana.

❿   Have the students answer the questions.

PHYSICS

Results

Have the students calculate the GPE for the toy car with and without the pennies at the corresponding heights at which the banana was smashed. The formula is:

gravitational potential energy (GPE) = weight x height

The GPEs should be roughly equal. Basically, we expect that it takes a given amount of KE (kinetic energy) to smash the banana, and it doesn't matter whether this comes in the form of a heavy, slow car or a light, fast car. The GPE the students calculate is the energy needed to smash the banana.

III. Conclusions

Have the students review the results they recorded for the experiment. Have them draw conclusions based on the data they collected.

IV. Why?

Read this section of the *Laboratory Notebook* with your students.
Discuss any questions that might come up.

V. Just For Fun

Have the students repeat the experiment using a raw egg instead of a banana. They should discover that it takes more force to break an egg than to smash a banana slice. The amount of force can be increased by raising the ramp and/or putting more weight on the car.

PHYSICS

Experiment 13

On Your Own

Materials Needed

- student-selected materials
- several sheets of paper

Objectives

In this unit students will create their own experiment to observe the law of conservation of energy as one form of energy converts to another.

The objectives of this lesson are for students to:

- Explore the scientific method by creating their own experiment.
- Observe energy converting from one form to another.

Experiment

I. Think About It

Read this section of the *Laboratory Notebook* with your students.

Ask questions such as the following to guide open inquiry.

- *What are some examples of energy being converted from one form to another?*

- *Do you think energy being converted from one form to another is important for life? Why or why not?*

- *Do you think the energy that comes from the Sun is important for life on Earth? Is it ever converted? Why or why not?*

- *Do you think any energy conversions happen when a car is started and driven away? Why or why not? What are they?*

- *Do you think energy is being converted when water goes down a waterfall? How?*

- *Do you think that as energy gets converted from one form to another there is less and less energy on Earth? Why or why not?*

II. Experiment 13: On Your Own

Students will create their own experiment to explore the concept of the law of conservation of energy by observing how one form of energy changes to another. The goal is to convert different energies into other forms of energy and to include as many different forms of energy as they can.

Review with the students the different types of energy conversions described in Chapters 12 and 13 of the *Student Textbook*. Have them read the introduction to the experiment.

PHYSICS

In the example given, the kinetic energy of a rolling marble is used to knock down a domino that has a cap filled with baking soda on top of it. The baking soda falls into vinegar and a chemical reaction occurs.

The marble begins with GPE which gets converted to KE as it rolls down the ramp. The KE is used to convert the GPE of the elevated baking soda into KE as it falls. This releases the chemical potential energy (CPE) in the baking soda and vinegar, and a chemical reaction begins, producing CO_2 which then puts out the fire. The chemical energy of the baking soda and vinegar is converted into heat energy and bubbles (gas). The gas rises and puts out the flame of a match.

Have the students think of ways this example might be changed. For example, the gas from the chemical reaction could be released into a small balloon or used to move a small piston.

❶-❷ Have students fill in the information requested.

❸ To prepare for the experiment, have the students do several "thought experiments" by asking themselves what different events they might include in their experiment that would change energy from one form to another. Some of their ideas will not be practical, but have them use their imagination to think of different ways to convert energy. Have them write down their ideas and identify what the energies will be before and after conversion.

❹ Once they have some different events in mind, they can think about ways to link the different ideas together in a series of events. Have them think about whether or not their ideas could work and what items they might use in their experiment.

The students will write an objective and hypothesis. Based on the series of events they have come up with, have them assemble a materials list, write down the steps of their experiment, make a drawing of their experimental setup. and then perform the experiment.

Have the students record their results whether or not their experiment worked. Have them write valid conclusions based on their results, and ask them what they might do differently next time.

III. Conclusions

Have the students review the results they recorded for the experiment. Have them draw conclusions based on the data they collected.

IV. Why?

Read this section of the *Laboratory Notebook* with your students.
Discuss any questions that might come up.

V. Just For Fun

Students will come up with a different setup for converting one form of energy to another. Encourage them to include more steps. They can use the same type of conversion as many times as they want and should also include some different types of conversions.

Experiment 14

Observing Your World

Materials Needed

- pencil, pen, colored pencils
- small jar
- trowel or spoon

Optional

- binoculars

Objectives

In this experiment students will collect data from their observations of the world around them and organize the data in a chart.

The objectives of this lesson are for students to:

- Make careful observations.
- Record and organize data.

Experiment

I. Think About It

Read this section of the *Laboratory Notebook* with your students.

Ask questions such as the following to guide open inquiry.

- *Do you think if you go outside and look at things carefully you will observe details that you hadn't noticed before? Why or why not?*

- *If you are riding your bike or walking outside, what kinds of things do you notice that tell you where you are?*

- *Do you think buildings are different in the city than they are in the country? Why or why not?*

- *Do you think there are the same kinds of landscapes in a city and in the country? Why or why not?*

- *What kinds of weather do you have in your area?*

- *Does weather affect the landscape where you live? Why or why not?*

II. Experiment 14: Observing Your World

In this experiment students will explore their local world—the world around them.

Have the students read the entire experiment, noting the experimental steps to be followed.

GEOLOGY

Objective: Have the students write an objective. Some examples:

> * *To learn more about the place where I live.*
>
> * *To better understand my hometown (or city, etc.)*
>
> * *To observe my surroundings and find out more about Earth through my observations.*

Hypothesis: Have the students write a hypothesis. Have them think about what they might learn by doing this experiment. Some examples:

> * *By making observations, I will better understand where I live.*
>
> * *By making observations, I can understand how the area where I live changes over time.*
>
> * *I will be able to assemble a list of features that describe where I live.*

❶-❸ Students walk to a place where they are standing on some type of ground (e.g., dirt, grass, concrete) and then observe what is around them. In the boxes provided, they will write or draw what they see. If binoculars are available, students can use them to observe the details of landforms and other features that are farther away.

❹-❺ Have students go to a place where they can dig up a small sample of dirt with a trowel or spoon. They can put the dirt sample in a small jar. Have them examine the dirt carefully, looking for color, texture, the presence of organic matter, and any other features they can observe. Have them record their observations.

❻ Have the students observe any man-made structures nearby. These can include things such as homes, roads, buildings, parks, fences, and utility poles.

❼ Have the students explore how dynamic features near their home may have changed the way the area looks. For example, heavy rains or snowmelt runoff might have caused erosion. Lightning might have damaged a tree.

❽ Have the students explore the history of the place where they live. Have them research when their house was built, when their city was founded, or when the land surrounding them became farms or industrial parks.

GEOLOGY

Results

Have the students use the chart to assemble the information they've collected.

III. Conclusions

Have the students review the results they recorded for the experiment. Have them draw conclusions based on the data they collected.

IV. Why?

Read this section of the *Laboratory Notebook* with your students.
Discuss any questions that might come up.

V. Just For Fun

Have the students imagine they got an email from someone on the planet Alpha Centauri Bb asking for information about what Earth is like. Have the students use the data they have collected to write a narrative describing the area where they live. Have them think about how they would describe the area so that someone who has not seen it could imagine it.

GEOLOGY

Experiment 15

Mineral Properties

Materials Needed

- known mineral samples:
  calcite
  feldspar
  quartz
  hematite
- several rocks from backyard
  or nearby
- copper penny
- steel nail
- streak plate (unglazed white
  ceramic tile)
- paper
- scissors
- marking pen
- tape
- vinegar (small amount)
- lemon juice (small amount)
- eyedropper or spoon

You can find minerals at a local
rock and mineral store or order
them online.

A Mineral Scale of Hardness Set
of minerals is available from Home
Science Tools
http://www.hometrainingtools.com

Objectives

In this experiment students will begin to explore the Earth by learning about minerals and rocks.

The objectives of this lesson are for students to:

- Use scientific processes for identifying minerals.
- Use data from charts and make their own charts.

Experiment

I. Think About It

Read this section of the *Laboratory Notebook* with your students.

Ask questions such as the following to guide open inquiry.

- *Do you think rocks stay the same or change over time? Why?*
- *Do you think all rocks are made the same way? Why or why not?*
- *Do you think dirt (soil) has anything to do with rocks? Why or why not?*
- *How do you think soil is made?*
- *What do you think rocks and minerals can be used for?*
- *Do you think there are ways to tell what kind of rock you have? Why or why not?*

II. Experiment 15: Mineral Properties

Have the students read the entire experiment before beginning.

Objective: Have the students write an objective. Some examples:

- *To determine what minerals are in my backyard.*
- *To test rocks for the minerals calcite, quartz, feldspar, and hematite.*
- *To learn how to use field tests to determine which minerals are in rocks.*

GEOLOGY

Hypothesis: Have the students write a hypothesis. The hypothesis should be specific and can be about their backyard rocks or the known minerals they test. Some examples:

- *Rocks in my backyard contain quartz.*

- *Rocks in my backyard do not contain quartz.*

- *Feldspar is harder than calcite.*

- *Quartz is softer than hematite.*

Experiment—Part I

▶ Using the Mohs hardness scale provided in the *Laboratory Notebook,* have the students test each of the four known minerals for hardness.

The way to use the scale is as follows:

- *If a fingernail is able to scratch the mineral, the hardness is below 2.5.*

- *If a fingernail is not able to scratch the mineral, but a copper penny does, then the hardness is between 2.5 and 3.*

- *If a copper penny cannot scratch the mineral but a steel nail can, then the hardness is between 3 and 5.5.*

- *If a steel nail cannot scratch the mineral but an unglazed ceramic plate can (the mineral leaves a "streak"), then the hardness is between 5.5. and 6.5.*

- *If the mineral is not scratched by the ceramic plate (does not leave a streak) then the mineral is harder than 6.5.*

Have the students record other observations, such as size, color, and texture of the mineral samples. Also, have the students note any crystalline properties. It is important for students to record what they actually observe even if it's not what they think the "right" answer would be.

Typical hardness values are:

- Calcite: 3
- Quartz: 7
- Feldspar: 6
- Hematite: 5-6

GEOLOGY

Experiment—Part II

▶ Using the ceramic streak plate, have the students record the color streak left by each known mineral. To determine the color streak, they will take each mineral and scrape it firmly across the streak plate.

The color left on the plate is the "streak color" of the mineral. The streak color can be different from the color of the mineral sample as a whole. If this occurs, have the students note any differences. Some minerals, such as calcite, have a white streak.

Have students record their results.

Experiment—Part III

❶ Have the students go to the backyard or a place nearby and collect several rocks that look different from each other in color and texture.

❷-❹ Have the students label with numbers the rocks they collected and then record the numbers in the chart provided.

Students will do a hardness test and a streak test for each rock. Have them record their results and then compare them to the data collected in Part I and Part II to see if they can determine whether any of the known minerals they tested can be found in the locally collected rocks. They may or may not be able to determine whether the minerals they tested are in the unknown rock samples.

Other minerals can be tested if desired.

III. Conclusions

Have the students review the results they recorded for the experiment. Have them draw conclusions based on the data they collected.

The second part of this section asks the students to consider the subjective aspects of the hardness test and the streak test. They are subjective in that one person might see a slightly different color or observe a different hardness for the same mineral than another person, or objects used for testing might vary in actual hardness. Have the students think about how the subjectivity of the test might change the results.

An objective test is one where the observers' opinions may matter less, though this is not always the case. Have the students discuss whether or not chemical analysis might be more objective and less subjective than either the hardness test or streak test.

GEOLOGY

IV. Why?

Read this section of the *Laboratory Notebook* with your students.
Discuss any questions that might come up.

V. Just For Fun

Students will test their rock and mineral samples first with vinegar and then with lemon juice to see if any of them will have a chemical reaction with these acids. Have them rinse the vinegar off the rocks before testing with lemon juice. Have them record their results.

Have the students compare their results to see if the rock samples reacted in the same way to lemon juice as they did to vinegar.

Students can look on the internet or at the library to find out how the known mineral samples would be expected to react with acid. If their results varied, discuss how acids can be different strengths and the strength of the acid could affect the outcome of their experiment.

Experiment 16

Model Earth

Materials Needed

- Students will select materials and use them to make a model of Earth's layers

Chocolate Lava Cake

- butter: 113 grams (1/2 cup)
- semi-sweet chocolate chips: 133 ml (1/2 cup + 1 Tbsp.)
- 2 whole eggs
- 2 egg yolks
- powdered sugar: 192 ml (3/4 cup + 1 Tbsp.)
- flour: 94 ml (1/3 cup + 1 Tbsp.)

Objectives

In this experiment students will explore Earth's layers by making a model of Earth.

The objectives of this lesson are for students to:

- Explore model making as a way to understand Earth's layers.
- Learn about the benefits and limitations of using models in scientific discovery.

Experiment

I. Think About It

Read this section of the *Laboratory Notebook* with your students.

Ask questions such as the following to guide open inquiry.

- *How far below the surface of the Earth do you think rocks, minerals, and dirt extend? Why?*

- *What else do you think is below the Earth's surface? Gas? Melted rocks? Water? Living things?*

- *What do you think is at the center of the Earth? Is the center hard? Soft? Both? Neither?*

- *Do you think it's important that Earth has layers? Why or why not?*

II. Experiment 16: Model Earth

In this experiment students will create a model of Earth using what they have learned in this chapter of the *Student Textbook*. Students will design a model based on their own ideas about what materials could be used to represent Earth's different layers.

Have the students read the entire experiment.

Objective: Have the students write an objective. Some examples:

- *To explore building a model of Earth.*

- *To understand how to create accurate models of Earth.*

- *To explore the benefits and limitations of building models.*

GEOLOGY

Hypothesis: Have the students write a hypothesis. Some examples:

> • *I can create a model of Earth that accurately represents Earth's layers.*
>
> • *I cannot create a model of Earth that accurately represents Earth's layers.*
>
> • *I can create a model of Earth that teaches me something about Earth's layers.*

❶ Have the students use the *Student Textbook*, the internet, and/or the library to collect information about Earth's layers. The more research they do, the more accurate the model they will be able to build and the more they will learn about Earth's layers.

In the chart provided have the students list the features and depths of the layers.

❷ Have students think about the characteristics of each layer and imagine materials they could use to create the layer so that the characteristics match.

Students can use clay, food items, Styrofoam, cloth, felt, dirt, or any other materials they think would work for assembling a model of Earth.

❸ Help students think about which layers and features they want to represent in their model. Encourage them to represent as many layers as possible. Have the students choose the model materials from their chart.

For example:

> • *Students might want to build a model that represents the roundness of the Earth. In this case modeling clay of different colors could be used because it is easy to mold. A rock or small Styrofoam ball could be used for the solid inner core, a different color clay for each layer, and cloth for the crust. The clay layers will all be the same consistency.*
>
> • *They might want to represent the rigid lithosphere and the soft asthenosphere. In this case a pan with layered food items could be used (For example, peanut brittle (lithosphere) on top of peanut butter (asthenosphere). Some hard candy or a chocolate bar might represent the inner core.*
>
> • *If they chose to use a Styrofoam ball, then there would be no layers.*

❹ Have the students assemble their model.

Results

❶ Have the students determine how well their model represents Earth's geology. The models are not going to be perfect, and in the process of model building some features will be more accurately represented than others.

❷ Have the students fill in the chart provided. There are no right answers and their answers will be based on what they actually observed.

III. Conclusions

Have the students review the results they recorded and evaluate their models. Help them explore how challenging model building can be and the difficulties that arise when trying to build accurate models.

IV. Why?

Read this section of the *Laboratory Notebook* with your students.
Discuss any questions that might come up.

V. Just For Fun

A fun way to model the Earth is to make a chocolate lava cake to represent the hot liquid of the Earth's outer core surrounded by the firmer mantle. Students can poke a walnut or cherry into the cake to represent Earth's inner core. They might add Earth's crust by sprinkling nuts over the top of the cake.

CHOCOLATE LAVA CAKE

butter   113 grams (1/2 cup)
semi-sweet chocolate chips   133 ml (1/2 cup + 1 Tbsp.)
2 whole eggs
2 egg yolks
powdered sugar   192 ml (3/4 cup + 1 Tbsp.)
flour   94 ml (1/3 cup + 1 Tbsp.)

Microwave butter briefly until melted. Stir in chocolate chips until melted. Mix in whole eggs and yolks, then powdered sugar. Stir in flour. Pour into custard cups thoroughly greased with butter. Bake at 190° C (375° F) until edges are set and centers are still soft, about 10-13 minutes. Be careful not to overbake.
Makes about 4.

Have the students evaluate their lava cake model and compare it to their first model.

GEOLOGY

Dynamic Earth

Materials Needed

- student-made brittle candy (see below)
- 1 jar smooth peanut butter (for students with allergies to peanuts, whipped cream can be substituted)
- 118 ml (1/2 cup) crushed graham crackers
- plate or second cookie sheet
- materials to make a model volcano—student's choice

Materials for making brittle candy

Ingredients

237 ml (1 cup) white sugar
118 ml (1/2 cup) light corn syrup
1.25 ml (1/4 teaspoon) salt
59 ml (1/4 cup) water
28 grams (2 Tbsp) butter, softened
5 ml (1 teaspoon) baking soda

Equipment

2 liter (2 qt) saucepan
candy thermometer
cookie sheet, approx.
   30x36 cm (12x14 inches)
2 spatulas

Objectives

In this experiment students will build models to explore Earth's dynamics.

The objectives of this lesson are for students to:

- Explore model making for geological study.
- Observe how different layers of the Earth work together.

Experiment

I. Think About It

Read this section of the *Laboratory Notebook* with your students.

Ask questions such as the following to guide open inquiry.

- *How do you think mountains are formed?*
- *What do you think causes earthquakes?*
- *What do you think causes volcanoes?*
- *What happens to the surrounding area when a volcano erupts?*
- *What causes tectonic plates to move?*

II. Experiment 17: Dynamic Earth

Students will build a model using food items to examine how the movement of Earth's tectonic plates causes earthquakes to occur and mountains to form.

Have the students read the entire experiment.

Objective: Have the students write an objective. Some examples:

- *To explore how Earth's plates move on the putty-like asthenosphere.*
- *To experiment with a model of the crust, lithosphere, and asthenosphere to learn about plate tectonics.*
- *To explore the benefits and limitations of building models.*

GEOLOGY

Hypothesis: Have the students write a hypothesis. Some examples:

> • *I can create a model that will help me understand how plate tectonics works.*
>
> • *Making a model will show me something I did not expect about plate tectonics.*

❶ Students will make brittle candy that will represent tectonic plates in their model.

Help the students follow the recipe to make the brittle candy. It can be tricky to determine when the candy has cooked long enough. Make sure the students stir and check the candy often, noting the temperature and using fresh cold water each time they test the candy.

❷-❺ Once the candy has cooled, have the students create a model of the three layers—asthenosphere, lithosphere, and crust—using peanut butter, brittle candy, and the peanut butter/graham cracker mixture.

If a student has allergies to peanuts, whipped cream can be used in place of peanut butter.

❻ In the chart provided, have the students write the layer of Earth that each food item represents.

❼ Have the students move the brittle candy pieces on top of the peanut butter.

Have them bump the pieces together, scrape them alongside each other, and move the pieces up and down with respect to each other.

The students should pay special attention to the graham cracker/peanut butter topping. This layer represents the crust and is where the most visible changes will occur.

Have the students explore what happens when the brittle candy pieces are moved slowly, quickly, and collide or move past each other.

Explore with the students how well this works as a model of plate tectonics.

❽ Have the students write their observations in the box provided.

GEOLOGY

Results

Have the students review their observations and use them in recording their answers to the questions in the chart provided.

Have the students discuss the ways in which their model reflects the theory of plate tectonics and how it does not. For example:

- *The textures of the three foods may adequately represent the differences between the crust, lithosphere, and asthenosphere.*

- *The peanut butter may be pushed up to the surface where the "crust" is. This does not appear to happen in reality.*

- *The graham cracker/peanut butter crust may slide too easily off the brittle candy pieces, which doesn't reflect reality.*

- *The graham cracker/peanut butter crust mounds up like mountains.*

III. Conclusions

Have the students review the observations and results they recorded for the experiment and use them to evaluate their model.

Help the students explore how challenging model building can be and the difficulties that arise when trying to build accurate models.

IV. Why?

Read this section of the *Laboratory Notebook* with your students.
Discuss any questions that might come up.

V. Just For Fun

Have the students do online or library research to find a model volcano they can build. Once they have decided which model they want to make, they will create their own experiment, following the format of previous experiments.

Constellations

Materials Needed

- pencil
- flashlight
- compass

A clear night sky away from bright lights is needed.

Objectives

In this experiment students will become more familiar with the night sky, finding constellations, and navigating by the stars.

The objectives of this lesson are to have students:

- Locate a constellation.
- Observe how stars can be used for navigation.

Experiment

I. Think About It

Read this section of the *Laboratory Notebook* with your students.

Ask questions such as the following to guide open inquiry.

- *Before the compass was invented, do you think people were able to tell which direction they were going in at night? Why or why not?*

- *Do you think you could tell which way you are going at night if you don't have a compass? Why or why not?*

- *Do you think astronomers today keep making up new constellations? Why or why not?*

- *Do the constellations stay in the same position in the sky all night long? Why or why not?*

- *Do you think knowing the constellations could help you tell what time of night it is? Why or why not?*

II. Experiment 18: Constellations

This experiment has a section for the Northern Hemisphere and one for the Southern Hemisphere. At different times of year, students close to the equator may be able to observe all the constellations in this experiment.

Students will need to be away from bright lights to see the stars clearly.

Have the students read the entire experiment.

ASTRONOMY

Objective: Have the students write an objective. Some examples:

> • *To find a constellation.*
>
> • *To find north (or south) by using the stars.*

Hypothesis: Have the students write a hypothesis. Some examples:

> • *Finding a constellation will help me learn more about the stars.*
>
> • *Finding the North Star (or Southern Cross) will help me figure out where I am at night.*

EXPERIMENT

Have the students fill in the table provided with the information requested about their location.

Northern Hemisphere

❶-❷ Take the students outside and have them bring along a flashlight to use while reading and making drawings and notes. Students will follow the instructions in the *Laboratory Notebook* to find the Big Dipper, the North Star (Polaris), and the Little Dipper.

❸ Have the students draw the Little Dipper as they observe it.

❹-❻ Students will follow the instructions in the *Laboratory Notebook* to find the Dragon constellation, draw it, and answer the questions.

❼ Have the students locate the North Star again and use the compass to verify that the North Star is above the North Pole.

Southern Hemisphere

❶-❸ Take the students outside and have them bring along a flashlight to use while reading and making drawings and notes. Students will locate the Southern Cross and follow the instructions in the *Laboratory Notebook* to find south. Since the Southern Cross is not directly over the South Pole, students will attempt to locate the South Celestial Pole, which is the point in the sky directly above the South Pole. There is no pole star in the Southern Hemisphere.

❹ Students will lower their eyes from the South Celestial Pole straight down to the horizon. Have them note a landmark at that spot.

ASTRONOMY

❺ Have the students point their compass at the landmark they've noted and see if they have found south by using the Southern Cross as a pointer. They probably will not have accurately found south but should observe that by using this method they found a direction that is close to south.

❻ Have the students record their results.

❼ Have the students locate both the Southern Cross and the nearby False Cross and draw what they see.

III. Conclusions

Have the students review the results they recorded for the experiment. Have them answer the questions, drawing conclusions based on the data they collected.

IV. Why?

Read this section of the *Laboratory Notebook* with your students.
Discuss any questions that might come up.

V. Just For Fun

Have the students look online or in the library to find three additional constellations that can be seen from their area and that they would like to look for in the sky. Have them go outside and locate the constellations and then draw them as they see them.

ASTRONOMY

Experiment 19

Lunar and Solar Eclipses

Materials Needed

- basketball
- ping-pong ball
- flashlight
- empty toilet paper tube
- tape
- scissors
- a dark room
- student-selected objects

Objectives

By modeling lunar and solar eclipses, students will explore Earth and its interaction with the Moon and the Sun.

The objectives of this lesson are to have students:

- Learn about the relative positions and orbits of the Earth, Sun, and Moon.
- Observe how using a model can help with understanding Earth and its position in space.

Experiment

I. Think About It

Read this section of the *Laboratory Notebook* with your students.

Ask questions such as the following to guide open inquiry.

- *Do you think you think a lunar eclipse happens every time there is a full moon? Why or why not?*

- *If a solar eclipse is happening, do you think it can be seen from anywhere on Earth? Why or why not?*

- *Do you think lunar eclipses tell you anything about the Earth and the Moon? Why or why not?*

- *What do you think a lunar eclipse would look like if Earth were square?*

- *What do you think a solar eclipse would look like if the Moon were square?*

II. Experiment 19: Lunar and Solar Eclipses

In this experiment students will use a basketball, ping-pong ball, and flashlight to model how lunar and solar eclipses occur.

Have the students read the entire experiment.

Objective: Have the students write an objective. For example:

- To examine the difference between lunar and solar eclipses.

Hypothesis: Have the students write a hypothesis. For example:

- The ping-pong ball must be in direct alignment with the flashlight to make a shadow on the basketball.

☆☆
☆○ **ASTRONOMY**
○○
☆

EXPERIMENT

The objective of the experiment is for the students to see how a shadow is cast by an object passing between either the Moon and the Sun or the Earth and the Sun. This exercise can be performed by more than one student at a time, with the basketball, flashlight, and ping-pong ball each held by a student.

❶ In a dark room, students will place the basketball on one end of an upright empty toilet paper tube.

❷ Have students shine a flashlight on the basketball from several feet away. Then they will place the flashlight on the floor so that it still shines on the basketball.

❸ Have the students hold the ping-pong ball between the flashlight and the basketball so that a shadow is cast on the basketball.

It may take some time to find the location of the ping-pong ball that will result in a shadow falling on the basketball. Have the students note where they must place the ping-pong ball in respect to the flashlight in order to create a shadow. The ping-pong ball cannot be too high or too low, but needs to be in direct alignment with the flashlight.

❹-❺ Have the students move the ping-pong ball upward and then downward until no shadow is cast on the basketball in either position.

❻ Students will now move the ping-pong ball around the basketball to model the orbit of the Moon around the Earth. Have them observe where the ping-pong ball needs to be in order for the basketball to cast a shadow on the ping-pong ball and where the ping-pong ball needs to be to cast a shadow on the basketball. Also have them note the position of the ping-pong ball when no shadows are cast.

Results

Have the students repeat Step ❻ several times, making different orbits. Have them draw and label the results of several orbits, noting the position of the balls and where and whether shadows were cast. An example follows.

ASTRONOMY

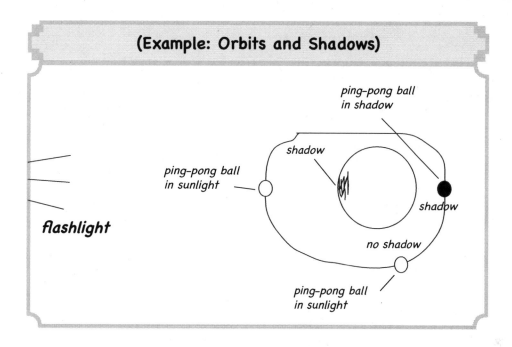

(Example: Orbits and Shadows)

ping-pong ball
in shadow

shadow

ping-pong ball
in sunlight

shadow

flashlight

no shadow

ping-pong ball
in sunlight

III. Conclusions

Have the students review the results they recorded for the experiment. Have them draw conclusions based on the data they collected.

IV. Why?

Read this section of the *Laboratory Notebook* with your students.
Discuss any questions that might come up.

V. Just For Fun

Students can do both **Part A** and **Part B** or choose one.

Part A: Students will use the same experimental setup to model how day and night occur on the Moon. Have the students put a mark on the ping-pong ball to identify the side of the Moon that always faces Earth and have them observe how light travels around the surface of the "Moon," creating day and night.

Part B: Students are to think of different configurations of planet(s), moon(s), and sun(s) that might occur in different solar systems. They can either use their experimental setup to make models or do this as a thought experiment with words and drawings. Encourage them to come up with changes of their own. Let them use their imagination freely even if you know their ideas are improbable. Have them record their results.

ASTRONOMY

Experiment 20

Modeling the Moon

Materials Needed

- modeling clay in the
  following colors
    gray
    white
    brown
    red
- butter knife or sculptor's knife
- ruler

Materials other than clay can be
used, such as Styrofoam balls or
plaster of Paris and paint.

Objectives

In this experiment students will build a model to explore features of the Moon.

The objectives of this lesson are to have students:

- Explore the structure of the Moon by building a model.
- Observe how scientists gain understanding by building models.

Experiment

I. Think About It

Read this section of the *Laboratory Notebook* with your students.

Ask questions such as the following to guide open inquiry.

> - *What do you think the Moon is made of?*
>
> - *What do you think the surface of the Moon is like?*
>
> - *Do you think you could see any features of the Moon's surface if you looked at it through binoculars? Why or why not?*
>
> - *What do you think it would be like to travel on the surface of the Moon? Would you be able to breathe Moon air? Would it be hard to jump up? Would you find mud?*
>
> - *Do you think people could live on the Moon? How do you think they would do it?*

II. Experiment 20: Modeling the Moon

Have the students read the entire experiment.

Objective: Have the students write an objective. Some examples:

> - *In this experiment we will explore features of the Moon by creating a clay model.*
>
> - *Models play an important role in science, and we will use modeling to explore the Moon.*
>
> - *We will use models to explore the three-dimensional nature of the Moon.*

☆☆ **ASTRONOMY**
☆○
☆☆

Hypothesis: Have the students write a hypothesis. Some examples:

> • *By building a model of the Moon, we will get an idea of what the Moon looks like in 3D.*
>
> • *The Moon model will help us learn about the features of the Moon.*

EXPERIMENT

Model building is an important part of science. Models help scientists visualize how something might look in three dimensions.

❶ Have the students carefully observe the cutaway image of the Moon in the *Student Textbook*. Have them notice details.

❷ Have the students build a model Moon based on the cutaway image. Have them make the model as accurate as possible.

❸ Have the students use a ruler to measure the diameter of their finished model.

Results

(Answers will vary. The answers shown are an example only.)

The real Moon is 3476.2 kilometers (2160 miles) in diameter. Compare the diameter of your model with the actual diameter of the Moon. Do the following steps to calculate how many times smaller your model is compared to the actual size of the Moon.

❶ Write the diameter of your model Moon in centimeters _____ or in inches *5 inches* . The diameter of the actual Moon is 3476.2 kilometers (2160 miles).

❷ Convert the diameter of your model Moon to kilometers.

(If you are using inches, first multiply by 2.54 to get centimeters.

___5___ inches x 2.54 = __*12.7*__ centimeters.)

Multiply the number of centimeters by 0.00001 to get kilometers. This will be a very small number.

__*12.7*__ centimeters X 0.00001 = __*0.000127*__ kilometers.

❸ Divide the actual diameter of the Moon by the diameter of your model Moon.

3476.2 kilometers (actual Moon) ÷ __*0.000127*__ kilometers (model Moon) = *27,371,654*

This should be a very large number. It tells you how many times larger the real Moon is compared to your model Moon.

III. Conclusions

Have the students review the results they recorded for the experiment. Have them draw conclusions based on the data they collected.

IV. Why?

Read this section of the *Laboratory Notebook* with your students.
Discuss any questions that might come up.

V. Just For Fun

Students are to imagine they are designing a station on the Moon. Have them review what they have learned about the Moon and think about what people would need to live and work on the Moon. Have them record their ideas and then make a drawing of their Moon station. Encourage them to use their imagination freely even if you know their ideas wouldn't work.

Experiment 21

Modeling the Planets

Materials Needed

- modeling clay in the following colors:
    - gray
    - white
    - brown
    - red
    - blue
    - green
    - orange
- butter knife or sculptor's knife
- colored pencils

Objectives

In this experiment students will use model building to compare the features of the planets in our solar system.

The objectives of this lesson are to have students:

- Explore the features of both the terrestrial and Jovian planets.
- Learn more about the benefits and difficulties of model building in scientific exploration.

Experiment

I. Think About It

Read this section of the *Laboratory Notebook* with your students.

Ask questions such as the following to guide open inquiry.

- *What unique features do you think Earth has?*
- *Do you think Earth is similar to Mars? Why or why not?*
- *Do you think Earth is similar to Jupiter? Why or why not?*
- *What do you know about Jupiter, Saturn, and Uranus?*
- *What do you think Saturn's rings are made of?*
- *Do you think building models of the planets can help you learn more about their similarities and differences? Why or why not?*
- *Do you think scientists need to do research before building models? Why or why not?*
- *Do you think building models can help scientists find new questions to be researched? Why or why not?*

II. Experiment 21: Modeling the Planets

In this experiment students will build models of the eight planets of our solar system. To find out more about the planets, students can use additional references. Images of all the planets can be found on the internet by doing a search on the planet name.

ASTRONOMY

Have the students read the entire experiment.

Objective: Have the students write an objective. Some examples:

> - *In this experiment we will explore features of the planets by creating clay models.*
>
> - *Models play an important role in science, and we will use modeling to explore the planets.*
>
> - *We will use models to explore the three-dimensional nature of the planets.*

Hypothesis:: Have the students write a hypothesis. Some examples:

> - *By building models of the planets, we will get an idea of what the planets look like in 3D.*
>
> - *The models of the planets will help us learn about the features of the planets.*

EXPERIMENT

❶ Have the students look closely at the images of the planets in the *Student Textbook,* particularly their relative sizes, shape, colors, and surface patterns.

❷ Using the images in the textbook for reference, students are to make written notes about what they observe about each planet's features. Then they will make a colored sketch of each planet showing features they will be modeling. Spaces are provided.

Students can do online or library research to find more planet images. They can do a browser search for individual planet images or use websites such as www.nasa.gov/.

❸ Students will build their models while using their notes and sketches for reference. Remind them to keep the relative sizes of the planets in proportion.

Results

Have the students observe their models and record their observations.

ASTRONOMY

III. Conclusions

Have the students review the results they recorded for the experiment. Have them draw conclusions based on the data they collected.

IV. Why?

Read this section of the *Laboratory Notebook* with your students.
Discuss any questions that might come up.

V. Just For Fun

Students will imagine they are using a space telescope to view a planet for the first time. Have them name the planet and list features they imagine it to have. Students can review the notes they made for the planets in our solar system to get ideas, and they can also make up new features. Once they have listed the features, have them make a colored sketch of the planet. If they'd like to, they can build a model of it.

Encourage students to use their imagination freely. Using the imagination is an important part of doing science. There are no "right" answers to this experiment.

Experiment 22

Cameras Rolling!

Materials Needed

- a video recording device
  (camcorder, iPad, cell phone)

Objectives

In this experiment students will explore science by making a film.

The objectives of this lesson are for students to:

- Write a narrative about science.
- Research scientific facts and discoveries.

Experiment

I. Think About It

Read this section of the *Laboratory Notebook* with your students.

Ask questions such as the following to guide open inquiry.

- *What movies or TV programs have you watched that had science as a theme? Did you learn anything about science from them?*

- *What did the movie have to say about science and scientists?*

- *Do you think science and scientists were accurately protrayed? Why or why not?*

- *How interesting was the movie? Why?*

- *If you were to make a movie about science, what theme would you choose? Why?*

- *Do you think you would have to do a lot of research to make a movie that had a scientific theme? Why or why not?*

- *How could you include more than one science subject in your movie?*

II. Experiment 22: Cameras Rolling!

In this experiment students will make a video about science. Their video can be very simple or more complicated depending on the students' interest and the equipment available.

Have the students read the entire experiment.

Objective: Have the students write an objective.
Hypothesis: Have the students write a hypothesis.

EXPERIMENT

❶ Have the students review Chapter 22 of the *Student Textbook* and decide on an area of science, a scientist, or a scientific discovery they'd like to explore in a movie.

❷ Have the students decide what the movie will be about and what type of movie it will be. If they choose science fiction, the movie must be based on real science and scientists must be portrayed realistically (not mad scientists). They can imagine how science will have advanced in the future, but the foundation must be real science.

❸ Have the students write an outline for their story, including the characters that will be in it. Encourage them to do more research in previous chapters of the textbook, online, or at the library.

❹ Have the students write the script including dialog and notes about what the characters will be doing at different points in the movie. Have them think about the location and props, if needed.

❺ Have them recruit friends and family to be actors in the movie.

❻ Have them make the video.

Results

Have the students present their video to friends and family.

III. Conclusions

Have the students review their movie making process and draw conclusions about making scientific movies.

IV. Why?

Read this section of the *Laboratory Notebook* with your students.
Discuss any questions that might come up.

V. Just For Fun

Students can watch a movie based on science, evaluate it, and record their observations.

- Or -

Students will evaluate their own film to see what they could change about it to make it better. Then they will think of ideas for more movies about science and write them down in the space provided.

Optional: Students can choose one of their ideas to make another science movie.